Copyright © 2011 XAMonline, Inc.
All rights reserved. No part of the material protected by this copyright notice may be reproduced or utilized in any form or by any means, electronic or mechanical, including photocopying, recording or by any information storage and retrievable system, without written permission from the copyright holder.

To obtain permission(s) to use the material from this work for any purpose including workshops or seminars, please submit a written request to:

> XAMonline, Inc.
> 25 First Street, Suite 106
> Cambridge, MA 02141
> Toll Free: 1-800-509-4128
> Email: info@xamonline.com
> Web: www.xamonline.com
> Fax: 1-617-583-5552

Library of Congress Cataloging-in-Publication Data

Wynne, Sharon A.
  PRAXIS II Middle School Mathematics 0069 Practice Test 1: Teacher Certification /
  Sharon A. Wynne. -1st ed.
  ISBN: 978-1-60787-123-1
    1. PRAXIS II Middle School Mathematics 0069 Practice Test 1   2. Study Guides
    3. PRAXIS   4. Teachers' Certification & Licensure   5. Careers

Disclaimer:

The opinions expressed in this publication are the sole works of XAMonline and were created independently from the National Education Association, Educational Testing Service, or any State Department of Education, National Evaluation Systems or other testing affiliates.

Between the time of publication and printing, state specific standards as well as testing formats and website information may change that is not included in part or in whole within this product. Sample test questions are developed by XAMonline and reflect similar content as on real tests; however, they are not former tests. XAMonline assembles content that aligns with state standards but makes no claims nor guarantees teacher candidates a passing score. Numerical scores are determined by testing companies such as NES or ETS and then are compared with individual state standards. A passing score varies from state to state.

Printed in the United States of America   œ-1
PRAXIS II Middle School Mathematics 0069 Practice Test 1
ISBN: 978-1-60787-123-1

Praxis Middle School Mathematics 0069
Pre-Test Sample Questions

DIRECTIONS: Read each item and select the best response.

1. $24 - 3 \times 7 + 2 =$
   (Average) (Skill 1.2)

   A. 5

   B. 149

   C. −3

   D. 189

2. Which of the following sets is closed under division?
   (Average) (Skill 1.3)

   A. Integers

   B. Rational numbers

   C. Natural numbers

   D. Whole numbers

3. Joe reads 20 words/min., and Jan reads 80 words/min. How many minutes will it take Joe to read the same number of words that it takes Jan 40 minutes to read?
   (Rigorous) (Skill 1.5)

   A. 10

   B. 20

   C. 80

   D. 160

4. Choose the set in which the members are not equivalent.
   (Average) (Skill 1.6)

   A. 1/2, 0.5, 50%

   B. 10/5, 2.0, 200%

   C. 3/8, 0.385, 38.5%

   D. 7/10, 0.7, 70%

5. Given W = whole numbers, N = natural numbers

   Z = integers,
   R = rational numbers, and
   I = irrational numbers

   Which of the following is not true?
   (Easy) (Skill 1.7)

   A. $R \subset I$

   B. $W \subset Z$

   C. $Z \subset R$

   D. $N \subset W$

6. Which denotes a complex number?
   (Rigorous) (Skill 1.7)

   A. 3.678678678…

   B. $-\sqrt{27}$

   C. $123^{1/2}$

   D. $(-100)^{1/2}$

7. Express .0000456 in scientific notation.
   (Average) (Skill 1.8)

   A. $4.56x10^{-4}$

   B. $45.6x10^{-6}$

   C. $4.56x10^{-6}$

   D. $4.56x10^{-5}$

8. Find the GCF of $2^2 \cdot 3^2 \cdot 5$ and $2^2 \cdot 3 \cdot 7$.
   (Average) (Skill 1.11)

   A. $2^5 \cdot 3^3 \cdot 5 \cdot 7$

   B. $2 \cdot 3 \cdot 5 \cdot 7$

   C. $2^2 \cdot 3$

   D. $2^3 \cdot 3^2 \cdot 5 \cdot 7$

9. Which of the following is always composite if x is odd, y is even, and both x and y are greater than or equal to 2?
   (Rigorous) (Skill 1.11)

   A. $x+y$

   B. $3x+2y$

   C. $5xy$

   D. $5x+3y$

10. If three cups of concentrate are needed to make 2 gallons of fruit punch, how many cups are needed to make 5 gallons?
    (Easy) (Skill 1.12)

    A. 6 cups

    B. 7 cups

    C. 7.5 cups

    D. 10 cups

11. The volume of water flowing through a pipe varies directly with the square of the radius of the pipe. If the water flows at a rate of 80 liters per minute through a pipe with a radius of 4 cm, at what rate would water flow through a pipe with a radius of 3 cm?
    (Average) (Skill 1.12)

    A. 45 liters per minute

    B. 6.67 liters per minute

    C. 60 liters per minute

    D. 4.5 liters per minute

12. Simplify $\dfrac{\frac{3}{4}x^2y^{-3}}{\frac{2}{3}xy}$
    (Rigorous) (Skill 1.17)

    A. $\frac{1}{2}xy^{-4}$

    B. $\frac{1}{2}x^{-1}y^{-4}$

    C. $\frac{9}{8}xy^{-4}$

    D. $\frac{9}{8}xy^{-2}$

13. Solve: $\sqrt{75} + \sqrt{147} - \sqrt{48}$
    (Rigorous) (Skill 1.17)

    A. 174

    B. $12\sqrt{3}$

    C. $8\sqrt{3}$

    D. 74

14. Solve for x:
    $3x + 5 \geq 8 + 7x$
    (Average) (Skill 1.19)

    A. $x \geq -\frac{3}{4}$

    B. $x \leq -\frac{3}{4}$

    C. $x \geq \frac{3}{4}$

    D. $x \leq \frac{3}{4}$

15. What is the solution set for the following equations?
    (Rigorous) (Skill 1.19)
    $3x + 2y = 12$
    $12x + 8y = 15$

    A. All real numbers

    B. $x = 4, y = 4$

    C. $x = 2, y = -1$

    D. $\varnothing$

16. **Identify the proper sequencing of subskills when teaching graphing inequalities in two dimensions**
    (Easy) (Skill 1.19)

    A. Shading regions, graphing lines, graphing points, determining whether a line is solid or broken

    B. Graphing points, graphing lines, determining whether a line is solid or broken, shading regions

    C. Graphing points, shading regions, determining whether a line is solid or broken, graphing lines

    D. Graphing lines, determining whether a line is solid or broken, graphing points, shading regions

17. **Factor completely:**
    $8(x - y) + a(y - x)$
    (Average) (Skill 1.20)

    A. $(8 + a)(y - x)$

    B. $(8 - a)(y - x)$

    C. $(a - 8)(y - x)$

    D. $(a - 8)(y + x)$

18. **What would be the seventh term of the expanded binomial $(2a+b)^8$?**
    (Rigorous) (Skill 1.20)

    A. $2ab^7$

    B. $41a^4b^4$

    C. $112a^2b^6$

    D. $16ab^7$

19. **Solve for x:**
    $|2x+3| > 4$
    (Rigorous) (Skill 1.21)

    A. $-\frac{7}{2} > x > \frac{1}{2}$

    B. $-\frac{1}{2} > x > \frac{7}{2}$

    C. $x < \frac{7}{2}$ or $x < -\frac{1}{2}$

    D. $x < -\frac{7}{2}$ or $x > \frac{1}{2}$

20. **Given a 30-by-60-meter garden with a circular fountain with a 5 meter radius, calculate the area of the portion of the garden not occupied by the fountain.**
    (Rigorous) (Skill 2.2)

    A. 1721 m²

    B. 1879 m²

    C. 2585 m²

    D. 1015 m²

21. **The above diagram is most likely used in deriving a formula for which of the following?**
    (Easy) (Skill 2.2)

    A. The area of a rectangle

    B. The area of a triangle

    C. The perimeter of a triangle

    D. The surface area of a prism

22. **Determine the volume of a sphere to the nearest cm if the surface area is 113 cm².**
    (Rigorous) (Skill 2.2)

    A. 113 cm³

    B. 339 cm³

    C. 37.7 cm³

    D. 226 cm3

23. **Which is a postulate?**
    (Easy) (Skill 2.5)

    A. The sum of the angles in any triangle is 180°

    B. A line intersects a plane in one point

    C. Two intersecting lines form congruent vertical angles

    D. Any segment is congruent to itself

24. **Given similar polygons with corresponding sides of lengths 9 and 15, find the perimeter of the smaller polygon if the perimeter of the larger polygon is 150 units.**
    (Rigorous) (Skill 2.5)

    A. 54

    B. 135

    C. 90

    D. 126

25. Given $l_1 \parallel l_2$ prove $\angle b \cong \angle e$

   1) $\angle b \cong \angle d$ 1) vertical angle theorem

   2) $\angle d \cong \angle e$ 2) alternate interior angle theorem

   3) $\angle b \cong \angle e$ 3) symmetric axiom of equality

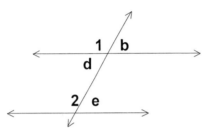

Which step is incorrectly justified?
(Average) (Skill 2.6)

A. Step 1

B. Step 2

C. Step 3

D. No error

26. What method could be used to prove the above triangles congruent?
(Average) (Skill 2.7)

A. SSS

B. SAS

C. AAS

D. SSA

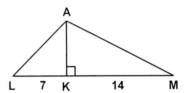

27. Given altitude AK in the right triangle ALM with measurements as indicated, determine the length of AK.
(Rigorous) (Skill 2.7)

A. 98

B. $7\sqrt{2}$

C. $\sqrt{21}$

D. $7\sqrt{3}$

28. Line *p* has a negative slope and passes through the point (0, 0). If line *q* is perpendicular to line *p*, which of the following must be true?
    (Rigorous) (Skill 2.12)

    A. Line *q* has a negative *y*-intercept

    B. Line *q* passes through the point (0,0)

    C. Line *q* has a positive slope

    D. Line *q* has a positive *y*-intercept

29. Find the distance between (3,7) and (−3,4).
    (Average) (Skill 2.12)

    A. 9

    B. 45

    C. $3\sqrt{5}$

    D. $5\sqrt{3}$

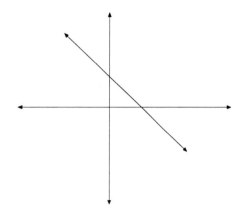

30. Which equation is represented by the above graph?
    (Average) (Skill 3.3)

    A. $x - y = 3$

    B. $x - y = -3$

    C. $x + y = 3$

    D. $x + y = -3$

31. Give the domain for the function over the set of real numbers:

    $$y = \frac{3x+2}{2x^2-3}$$

    (Rigorous) (Skill 3.7)

    A. All real numbers

    B. All real numbers, $x \neq 0$

    C. All real numbers, $x \neq -2$ or 3

    D. All real numbers, $x \neq \frac{\pm\sqrt{6}}{2}$

32. Which statement is true about George's budget?
    (Easy) (Skill 4A.2)

    A. George spends the greatest portion of his income on food

    B. George spends twice as much on utilities as he does on his mortgage

    C. George spends twice as much on utilities as he does on food

    D. George spends the same amount on food and utilities as he does on mortgage

33. Given a spinner with the numbers one through eight, what is the probability that you will spin an even number or a number greater than four?
    (Easy) (Skill 4A.3)

    A. 1/4

    B. 1/2

    C. 3/4

    D. 1

34. Given a drawer with 5 black socks, 3 blue socks, and 2 red socks, what is the probability that you will draw two black socks in two draws in a dark room?
    (Average) (Skill 4A.3)

    A. 2/9

    B. 1/4

    C. 17/18

    D. 1/18

35. If a horse will probably win three races out of ten, what are the odds that he will win?
    (Rigorous) (Skill 4A.3)

    A. 3:10

    B. 7:10

    C. 3:7

    D. 7:3

36. Find the median of the following set of data:

    14   3   7   6   11   20

    (Average) (Skill 4A.7)

    A. 9

    B. 8.5

    C. 7

    D. 11

37. **Corporate salaries are listed for several employees. Which would be the best measure of central tendency?**
    (Average) (Skill 4A.7)

    $24,000 $24,000 $26,000
    $28,000 $30,000 $120,000

    A. Mean

    B. Median

    C. Mode

    D. No difference

38. **A student scored in the 87th percentile on a standardized test. Which would be the best interpretation of his score?**
    (Easy) (Skill 4A.8)

    A. Only 13% of the students who took the test scored higher

    B. This student should be getting mostly Bs on his report card

    C. This student performed below average on the test

    D. This is the equivalent of missing 13 questions on a 100 question exam

39. **Determine the number of subsets of set K.**

    $K = \{4, 5, 6, 7\}$

    (Average) (Skill 4B.2)

    A. 15

    B. 16

    C. 17

    D. 18

40. $\{6, 11, 16, 21...\}$

    Find the sum of the first 20 terms in the sequence.
    (Rigorous) (Skill 4B.4)

    A. 1070

    B. 1176

    C. 969

    D. 1069

Constructed Response Questions

1. Three added to one-fourth of a number results in one-third of that number. Find the number.

    (a) Demonstrate at least two ways of solving this problem.
    (b) For each method, explain the steps.
    (c) For each method, identify the underlying concepts.

2. An open swimming pool 50 ft long and 20 ft wide has a wide walkway of uniform breadth running all around it. If the area of the walkway is four times that of the pool, what is the outer perimeter of the walkway? Round your answer to the tenth of a foot.

    (a) Draw and label a diagram to represent the problem described above.
    (b) Write an equation expressing the relationships between the different dimensions.
    (c) Explain the steps in your solution.

    How does the total area of the large rectangle including the pool and walkway change as the width of the walkway changes? Describe the pattern.

3. The frequency distribution below shows the summary of some test results where people scored points ranging from 0 to 10. One person scored 1 point, 4 people scored 2 points and so on.

    (a) Find the mean, median, mode and standard deviation of the data set.
    (b) Draw a histogram of the distribution.
    (c) How does the data compare to a normal distribution?

Points	Frequency	Cumulative Frequency
1	1	1
2	4	5
3	30	35
4	25	60
5	12	72
6	7	79
7	6	85
8	6	91
9	2	93
10	1	94

Praxis Middle School Mathematics 0069
Pre-Test Sample Questions with Rationales

The following represent one way to solve the problems and obtain a correct answer. There are many other mathematically correct ways of determining the correct answer.

1. $24 - 3 \times 7 + 2 =$
   (Average) (Skill 1.2)

   A. 5

   B. 149

   C. -3

   D. 189

Answer: A. 5
According to the order of operations, multiplication is performed first, and then addition and subtraction from left to right.

2. Which of the following sets is closed under division?
   (Average) (Skill 1.3)

   A. Integers

   B. Rational numbers

   C. Natural numbers

   D. Whole numbers

Answer: B. Rational numbers
In order to be closed under division, when any two members of the set are divided, the answer must be contained in the set. This is not true for integers, natural, or whole numbers as illustrated by the counter example 11/2 = 5.5.

3. Joe reads 20 words/min., and Jan reads 80 words/min. How many minutes will it take Joe to read the same number of words that it takes Jan 40 minutes to read?
   (Rigorous) (Skill 1.5)

   A. 10

   B. 20

   C. 80

   D. 160

Answer: D. 160
If Jan reads 80 words/minute, she will read 3200 words in 40 minutes.
$\frac{3200}{20} = 160$ At 20 words per minute, it will take Joe 160 minutes to read 3200 words.

4. Choose the set in which the members are <u>not</u> equivalent.
   (Average) (Skill 1.6)

   A. 1/2, 0.5, 50%

   B. 10/5, 2.0, 200%

   C. 3/8, 0.385, 38.5%

   D. 7/10, 0.7, 70%

Answer: C. 3/8, 0.385, 38.5%
3/8 is equivalent to .375 and 37.5%.

5. Given W = whole numbers, N = natural numbers,

    Z = integers,
    R = rational numbers, and
    I = irrational numbers

    Which of the following is not true?
    (Easy) (Skill 1.7)

    A. $R \subset I$

    B. $W \subset Z$

    C. $Z \subset R$

    D. $N \subset W$

Answer: A. $R \subset I$
The rational numbers are not a subset of the irrational numbers. All of the other statements are true.

6. **Which denotes a complex number?**
    (Rigorous) (Skill 1.7)

    A. 3.678678678...

    B. $-\sqrt{27}$

    C. $123^{1/2}$

    D. $(-100)^{1/2}$

Answer: D. $(-100)^{1/2}$
A complex number is the square root of a negative number. The complex number is defined as the square root of –1. The exponent ½ represents a square root.

7. **Express .0000456 in scientific notation.**
   (Average) (Skill 1.8)

   A. 4.56×10^{-4}

   B. 45.6×10^{-6}

   C. 4.56×10^{-6}

   D. 4.56×10^{-5}

Answer: D. 4.56×10^{-5}
In scientific notation, the decimal point belongs to the right of the 4, the first significant digit. To get from 4.56×10^{-5} back to 0.0000456, we would move the decimal point 5 places to the left.

8. **Find the GCF of $2^2 \cdot 3^2 \cdot 5$ and $2^2 \cdot 3 \cdot 7$.**
   (Average) (Skill 1.11)

   A. $2^5 \cdot 3^3 \cdot 5 \cdot 7$

   B. $2 \cdot 3 \cdot 5 \cdot 7$

   C. $2^2 \cdot 3$

   D. $2^3 \cdot 3^2 \cdot 5 \cdot 7$

Answer: C. $2^2 \cdot 3$
Choose the number of each prime factor that is in common.

9. Which of the following is always composite if x is odd, y is even, and both x and y are greater than or equal to 2?
   (Rigorous) (Skill 1.11)

   A. $x+y$

   B. $3x+2y$

   C. $5xy$

   D. $5x+3y$

Answer: C. $5xy$

A composite number is a number, which is not prime. The prime number sequence begins 2, 3, 5, 7, 11, 13, 17,…. To determine which of the expressions is <u>always</u> composite, experiment with different values of x and y, such as x=3 and y=2, or x=5 and y=2. It turns out that 5xy will always be an even number, and therefore, composite, if y=2.

10. If three cups of concentrate are needed to make 2 gallons of fruit punch, how many cups are needed to make 5 gallons?
    (Easy) (Skill 1.12)

    A. 6 cups

    B. 7 cups

    C. 7.5 cups

    D. 10 cups

Answer: C. 7.5 cups

Set up the proportion 3/2 = x/5, cross multiply to obtain 15=2x, then divide both sides by 2.

11. The volume of water flowing through a pipe varies directly with the square of the radius of the pipe. If the water flows at a rate of 80 liters per minute through a pipe with a radius of 4 cm, at what rate would water flow through a pipe with a radius of 3 cm?
    (Average) (Skill 1.12)

    A.   45 liters per minute

    B.   6.67 liters per minute

    C.   60 liters per minute

    D.   4.5 liters per minute

Answer: A. 45 liters per minute
Set up the direct variation: $\frac{V}{r^2} = \frac{V}{r^2}$. Substituting gives $\frac{80}{16} = \frac{V}{9}$. Solving for V gives 45 liters per minute.

12. Simplify $\dfrac{\frac{3}{4} x^2 y^{-3}}{\frac{2}{3} xy}$
    (Rigorous) (Skill 1.17)

    A. $\frac{1}{2} xy^{-4}$

    B. $\frac{1}{2} x^{-1} y^{-4}$

    C. $\frac{9}{8} xy^{-4}$

    D. $\frac{9}{8} xy^{-2}$

Answer: C. $\frac{9}{8} xy^{-4}$
Simplify the complex fraction by inverting the denominator and multiplying: 3/4(3/2) =9/8, then subtract exponents to obtain the correct answer.

13. Solve: $\sqrt{75} + \sqrt{147} - \sqrt{48}$
    (Rigorous) (Skill 1.17)

    A.   174

    B. $12\sqrt{3}$

    C. $8\sqrt{3}$

    D.   74

Answer: C. $8\sqrt{3}$

Simplify each radical by factoring out the perfect squares: $5\sqrt{3} + 7\sqrt{3} - 4\sqrt{3} = 8\sqrt{3}$

14. Solve for x:
    $3x + 5 \geq 8 + 7x$
    (Average) (Skill 1.19)

    A. $x \geq -\frac{3}{4}$

    B. $x \leq -\frac{3}{4}$

    C. $x \geq \frac{3}{4}$

    D. $x \leq \frac{3}{4}$

Answer: B. $x \leq -\frac{3}{4}$

Using additive equality, $-3 \geq 4x$. Divide both sides by 4 to obtain $-3/4 \geq x$. Carefully determine which answer choice is equivalent.

15. **What is the solution set for the following equations?**
    (Rigorous) (Skill 1.19)
    $3x + 2y = 12$
    $12x + 8y = 15$

    A. All real numbers

    B. $x = 4, y = 4$

    C. $x = 2, y = -1$

    D. $\varnothing$

Answer: D. $\varnothing$
Multiplying the top equation by −4 and adding results in the equation $0 = -33$. Since this is a false statement, the correct choice is the null set.

16. **Identify the proper sequencing of subskills when teaching graphing inequalities in two dimensions**
    (Easy) (Skill 1.19)

    A. Shading regions, graphing lines, graphing points, determining whether a line is solid or broken

    B. Graphing points, graphing lines, determining whether a line is solid or broken, shading regions

    C. Graphing points, shading regions, determining whether a line is solid or broken, graphing lines

    D. Graphing lines, determining whether a line is solid or broken, graphing points, shading regions

Answer: B. Graphing points, graphing lines, determining whether a line is solid or broken, shading regions

17. **Factor completely:**

    8(x – y) + a(y – x)

    (Average) (Skill 1.20)

    A. (8 + a) (y – x)

    B. (8 – a) (y – x)

    C. (a – 8) (y – x)

    D. (a – 8) (y + x)

Answer: C. (a – 8) (y – x)
Glancing first at the solution choices, factor (y – x) from each term. This leaves –8 from the first reran and a from the send term: (a – 8) (y – x)

18. **What would be the seventh term of the expanded binomial $(2a+b)^8$?**
    (Rigorous) (Skill 1.20)

    A. $2ab^7$

    B. $41a^4b^4$

    C. $112a^2b^6$

    D. $16ab^7$

Answer: C. $112a^2b^6$

The set-up for finding the seventh term is $\frac{8(7)(6)(5)(4)(3)}{6(5)(4)(3)(2)(1)}(2a)^{8-6}b^6$, which gives $28(4a^2b^6)$ or $112a^2b^6$. One can also find the right answer simply by looking at the powers of a and b.

19. Solve for x:
    $|2x+3| > 4$
    (Rigorous) (Skill 1.21)

    A. $-\frac{7}{2} > x > \frac{1}{2}$

    B. $-\frac{1}{2} > x > \frac{7}{2}$

    C. $x < \frac{7}{2}$ or $x < -\frac{1}{2}$

    D. $x < -\frac{7}{2}$ or $x > \frac{1}{2}$

Answer: D. $x < -\frac{7}{2}$ or $x > \frac{1}{2}$
The quantity within the absolute value symbols must be either > 4 or < –4. Solve the two inequalities 2x + 3 > 4 or 2x + 3 < –4

20. Given a 30-by-60-meter garden with a circular fountain with a 5 meter radius, calculate the area of the portion of the garden not occupied by the fountain.
    (Rigorous) (Skill 2.2)

    A. 1721 m²

    B. 1879 m²

    C. 2585 m²

    D. 1015 m²

Answer: A. 1721 m²
Find the area of the garden and then subtract the area of the fountain: $30(60) - \pi(5)^2$ or approximately 1721 square meters.

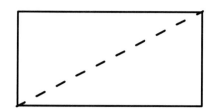

21. **The above diagram is most likely used in deriving a formula for which of the following?**
    (Easy) (Skill 2.2)

    A.   The area of a rectangle

    B.   The area of a triangle

    C.   The perimeter of a triangle

    D.   The surface area of a prism

Answer: B. The area of a triangle

22. **Determine the volume of a sphere to the nearest cm if the surface area is 113 cm².**
    (Rigorous) (Skill 2.2)

    A.   113 cm³

    B.   339 cm³

    C.   37.7 cm³

    D.   226 cm³

Answer: A. 113 cm³
Solve for the radius of the sphere using $A = 4\pi r^2$. The radius is 3. Then, find the volume using $4/3 \pi r^3$. Only when the radius is 3 are the volume and surface area equivalent.

23. **Which is a postulate?**
    (Easy) (Skill 2.5)

    A. The sum of the angles in any triangle is 180°.

    B. A line intersects a plane in one point.

    C. Two intersecting lines form congruent vertical angles.

    D. Any segment is congruent to itself.

Answer: D. Any segment is congruent to itself.
A postulate is an accepted property of real numbers or geometric figures, which cannot be proven; choices A, B. and C are theorems which can be proven.

24. **Given similar polygons with corresponding sides of lengths 9 and 15, find the perimeter of the smaller polygon if the perimeter of the larger polygon is 150 units.**
    (Rigorous) (Skill 2.5)

    A. 54

    B. 135

    C. 90

    D. 126

Answer: C. 90
The perimeters of similar polygons are directly proportional to the lengths of their sides, therefore 9/15 = x/150. Cross-multiply to obtain 1350 = 15x, then divide by 15 to obtain the perimeter of the smaller polygon.

25. Given $l_1 \parallel l_2$ prove $\angle b \cong \angle e$

   1) $\angle b \cong \angle d$ 1) vertical angle theorem
   2) $\angle d \cong \angle e$ 2) alternate interior angle theorem
   3) $\angle b \cong \angle 3$ 3) symmetric axiom of equality

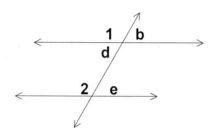

Which step is incorrectly justified?
(Average) (Skill 2.6)

A. Step 1

B. Step 2

C. Step 3

D. No error

Answer: C. Step 3
Step 3 can be justified by the transitive property.

26. **What method could be used to prove the above triangles congruent?**
    (Average) (Skill 2.7)

    A. SSS

    B. SAS

    C. AAS

    D. SSA

Answer: B. SAS
Use SAS with the last side being the vertical line common to both triangles.

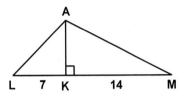

27. **Given altitude AK in the right triangle ALM with measurements as indicated, determine the length of AK.**
    (Rigorous) (Skill 2.7)

    A. 98

    B. $7\sqrt{2}$

    C. $\sqrt{21}$

    D. $7\sqrt{3}$

Answer: B. $7\sqrt{2}$
The attitude from the right angle to the hypotenuse of any right triangle is the geometric mean of the two segments that are formed. Multiply 7 x 14 and take the square root.

28. **Line p has a negative slope and passes through the point (0, 0). If line q is perpendicular to line p, which of the following must be true? (Rigorous) (Skill 2.12)**

    A. Line q has a negative y-intercept

    B. Line q passes through the point (0,0)

    C. Line q has a positive slope

    D. Line q has a positive y-intercept

Answer: C. Line q has a positive slope.
Draw a picture to help you visualize the problem.

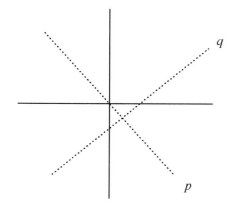

Choices A and D are not correct because line q could have a positive or a negative y-intercept. Choice B is incorrect because line q does not necessarily pass through (0, 0). Since line q is perpendicular to line p, which has a negative slope, it must have a positive slope.

29. Find the distance between (3,7) and (−3,4).
    (Average) (Skill 2.12)

    A.   9

    B.   45

    C. $3\sqrt{5}$

    D. $5\sqrt{3}$

Answer: C. $3\sqrt{5}$
Using the distance formula
$$\sqrt{[3-(-3)]^2 + (7-4)^2}$$
$$= \sqrt{36+9} = \sqrt{45} = 3\sqrt{5}$$

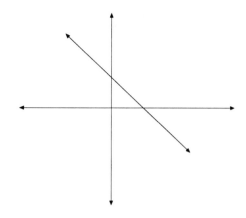

30. Which equation is represented by the above graph?
(*Average*) (*Skill 3.3*)

    A.   x − y = 3

    B.   x − y = −3

    C.   x + y = 3

    D.   x + y = −3

Answer: C. x + y = 3
By looking at the graph, we can determine the slope to be −1 and the y-intercept to be 3. Write the slope intercept form of the line as y = −1x + 3. Add x to both sides to obtain x + y = 3, the equation in standard form.

31. Give the domain for the function over the set of real numbers:

$$y = \frac{3x+2}{2x^2 - 3}$$

(Rigorous) (Skill 3.7)

- A. all real numbers
- B. all real numbers, $x \neq 0$
- C. all real numbers, $x \neq -2$ or 3
- D. all real numbers, $x \neq \dfrac{\pm\sqrt{6}}{2}$

Answer: D. all real numbers, $x \neq \dfrac{\pm\sqrt{6}}{2}$

Solve the denominator for 0. These values will be excluded from the domain.

$$2x^2 - 3 = 0$$
$$2x^2 = 3$$
$$x^2 = 3/2$$
$$x = \sqrt{\tfrac{3}{2}} = \sqrt{\tfrac{3}{2}} \cdot \sqrt{\tfrac{2}{2}} = \frac{\pm\sqrt{6}}{2}$$

32. **Which statement is true about George's budget?**
    (Easy) (Skill 4A.2)

    A. George spends the greatest portion of his income on food.

    B. George spends twice as much on utilities as he does on his mortgage.

    C. George spends twice as much on utilities as he does on food.

    D. George spends the same amount on food and utilities as he does on mortgage.

Answer: C: George spends twice as much on utilities as he does on food.

33. **Given a spinner with the numbers one through eight, what is the probability that you will spin an even number or a number greater than four?**
    (Easy) (Skill 4A.3)

    A. 1/4

    B. 1/2

    C. 3/4

    D. 1

Answer: C. 3/4
There are 6 favorable outcomes, 2, 4, 5, 6, 7, 8, and 8 possibilities. Reduce 6/8 to 3/4.

34. Given a drawer with 5 black socks, 3 blue socks, and 2 red socks, what is the probability that you will draw two black socks in two draws in a dark room?
    (Average) (Skill 4A.3)

    A.   2/9

    B.   1/4

    C.   17/18

    D.   1/18

Answer: A. 2/9
In this example of conditional probability, the probability of drawing a black sock on the first draw is 5/10. It is implied in the problem that there is no replacement, therefore the probability of obtaining a black sock in the second draw is 4/9. Multiply the two probabilities and reduce to lowest terms.

35. If a horse will probably win three races out of ten, what are the odds that he will win?
    (Rigorous) (Skill 4A.3)

    A.   3:10

    B.   7:10

    C.   3:7

    D.   7:3

Answer: C. 3:7
There are 3 slots to fill. There are 3 choices for the first, 7 for the second, and 6 for the third. Therefore, the total number of choices is 3(7)(6) = 126.

36. Find the median of the following set of data:

    14 3 7 6 11 20

    (*Average*) (*Skill 4A.7*)

    A. 9

    B. 8.5

    C. 7

    D. 11

Answer: A. 9
Place the numbers is ascending order: 3 6 7 11 14 20. Find the average of the middle two numbers: (7+11)/2 = 9.

37. Corporate salaries are listed for several employees. Which would be the best measure of central tendency?
    (*Average*) (*Skill 4A.7*)

    $24,000 $24,000 $26,000
    $28,000 $30,000 $120,000

    A. Mean

    B. Median

    C. Mode

    D. No difference

Answer: B. Median
The median provides the best measure of central tendency in this case where the mode is the lowest number and the mean would be disproportionately skewed by the outlier $120,000.

38. **A student scored in the 87th percentile on a standardized test. Which would be the best interpretation of his score?**
   (Easy) (Skill 4A.8)

   A. Only 13% of the students who took the test scored higher.

   B. This student should be getting mostly Bs on his report card.

   C. This student performed below average on the test.

   D. This is the equivalent of missing 13 questions on a 100-question exam.

Answer: A. Only 13% of the students who took the test scored higher.
Percentile ranking tells how the student compared to the norm or the other students taking the test. It does not correspond to the percentage answered correctly but can indicate how the student compared to the average student tested.

39. **Determine the number of subsets of set *K*.**

   K = {4, 5, 6, 7}

   (Average) (Skill 4B.2)

   A. 15

   B. 16

   C. 17

   D. 18

Answer: B. 16
A set of n objects has n^2 subsets. Therefore, here we have $4^2 = 16$ subsets. These subsets include four which each have 1 element only, six which each have 2 elements, four which each have 3 elements, plus the original set, and the empty set.

40. {6, 11, 16, 21...}

    Find the sum of the first 20 terms in the sequence.
    (Rigorous) (Skill 4B.4)

    A.   1070

    B.   1176

    C.   969

    D.   1069

Answer: A. 1070

Apply the formula $\frac{n}{2}[2a_1 + (n-1)d]$ where n = 20, a_1 =6 and d=5.

SAMPLE ANSWERS TO CONSTRUCTED RESPONSE QUESTIONS

1. **Three added to one-fourth of a number results in one-third of that number. Find the number.**

    (a) Demonstrate at least two ways of solving this problem.
    (b) For each method, explain the steps.
    (c) For each method, identify the underlying concepts.

Answer

First method: Three is the difference between one-third and one-fourth of the number.

Since $\frac{1}{3} - \frac{1}{4} = \frac{4-3}{12} = \frac{1}{12}$, $\frac{1}{12}$ of the number is 3.

Hence, the number is 3x12=36.

In order to solve the problem using this method, a student must be familiar with fractions and the standard method for addition and subtraction of fractions. The combination of fractions also requires the understanding of LCM.

Second method: This problem may be also solved using a diagrammatic representation of the two fractions involved.

A rectangle divided into 12 parts (the LCM of 3 and 4) is used to represent the two fractions of the number. From the diagram it is clear that one part in 12 is equal to 3. Hence the number is 12x3=36.

In order to use this method to solve the problem, a student must understand fractions as well s the concept of LCM.

Third method: This problem may also be solved using algebraic methods.
Let the unknown number be x. Then we can represent the given statement using the following equation:

$$\frac{1}{4}x + 3 = \frac{1}{3}x$$

Rearranging the terms and solving for x:

$$\frac{1}{3}x - \frac{1}{4}x = 3$$
$$\Rightarrow \frac{4x - 3x}{12} = 3$$
$$\Rightarrow \frac{x}{12} = 3$$
$$\Rightarrow x = 36$$

This method requires knowledge of basic algebra including the solution of equations using fractions.

2. An open swimming pool 50 ft long and 20 ft wide has a wide walkway of uniform breadth running all around it. If the area of the walkway is four times that of the pool, what is the outer perimeter of the walkway? Round your answer to the tenth of a foot.

   (a) Draw and label a diagram to represent the problem described above.
   (b) Write an equation expressing the relationships between the different dimensions.
   (c) Explain the steps in your solution.

   How does the total area of the large rectangle including the pool and walkway change as the width of the walkway changes? Describe the pattern.

Answer

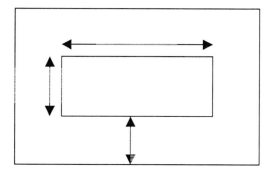

Let the width of the walkway be x ft.
Then, the length of the large rectangle = 50 + 2x ft
The width of the large rectangle = 20 + 2x ft

The area of the swimming pool = 20 x 50 = 1000 sq. ft
The area of the large rectangle including the walkway and swimming pool = (20 + 2x)(50 + 2x) sq. ft

Since the area of the walkway is 4 times that of the swimming pool, the area of the large rectangle including both the walkway and swimming pool is 5 times that of the swimming pool.

This relationship maybe expressed using the following quadratic equation:

$$(20+2x)(50+2x) = 5000$$

Multiplying out the terms and rearranging them in the standard form we get:

$$1000 + 100x + 40x + 4x^2 = 5000$$
$$\Rightarrow 4x^2 + 140x + 1000 = 5000$$
$$\Rightarrow 4x^2 + 140x - 4000 = 0$$
$$\Rightarrow x^2 + 35x - 1000 = 0$$

The quadratic formula can be used to solve the equation:

$$x = \frac{-35 \pm \sqrt{35^2 - 4 \cdot 1 \cdot (-1000)}}{2} = \frac{-35 \pm \sqrt{1225 + 4000}}{2} = \frac{-35 \pm 72.28}{2}$$

Since the width of the walkway cannot be a negative number, we choose the positive solution:

$$x = \frac{-35 + 72.28}{2} = \frac{37.28}{2} = 18.64$$

Hence, the length of the large rectangle = 50+18.64 x 2 = 50 + 37.28 = 87.3 ft (to the nearest tenth of a foot).

The width of the large rectangle = 20 + 37.28 = 57.3 ft
Thus, the perimeter of the large rectangle = 2(87.3+57.3)= 289.2 ft

The area of the large rectangle is related to the width of the walkway as follows:

Area = $(20 + 2x)(50 + 2x) = 4x^2 + 140x + 1000$

This is a quadratic function, and hence we expect to see the area increase in a non-linear fashion as x increases.

To observe the relation between the area of the large triangle and the value of x, we tabulate a set of values with a fixed increment:

X	Length of rectangle	Width of rectangle	Area
0	50	20	1000
5	60	30	1800
10	70	40	2800
15	80	50	4000
20	90	60	5400
25	100	70	7000

The table shows that every time the width of the walkway is increased by 5 ft, the area of the large rectangle increases by 200 sq. ft. more than it did the previous time.

3. The frequency distribution below shows the summary of some test results where people scored points ranging from 0 to 10. One person scored 1 point, 4 people scored 2 points and so on.

   (a) Find the mean, median, mode and standard deviation of the data set.
   (b) Draw a histogram of the distribution.
   (c) How does the data compare to a normal distribution?

Points	Frequency	Cumulative Frequency
1	1	1
2	4	5
3	30	35
4	25	60
5	12	72
6	7	79
7	6	85
8	6	91
9	2	93
10	1	94

Answer
The mean of the data set =

(1x1+2x4+3x30+4x25+5x12+6x7+7x6+8x6+9x2+10x1)= (1+8+90+100+60+42+42+48+18+10)=419/94=4.5

The median score (the point at which the cumulative frequency reaches or surpasses the value 47) = 4

The mode or value with the highest frequency = 3

The standard deviation of the data set =

$$\sqrt{\frac{(-3.5)^2 + 4(-2.5)^2 + 30(-1.5)^2 + 25(-0.5)^2 + 12(0.5)^2 + 7(1.5)^2 + 6(2.5)^2 + 6(3.5)^2 + 2(4.5)^2 + (5.5)^2}{94}}$$

$$\sqrt{\frac{12.25 + 25 + 67.5 + 62.5 + 3 + 15.75 + 37.5 + 73.5 + 40.5 + 30.25}{94}} = \sqrt{\frac{367.75}{94}} = 2.0$$

The frequency distribution from the above example is displayed below as a histogram.

This distribution is not normal. A normal distribution is symmetric with one peak exactly in the middle. Also, the mean, median, and mode are identical for a normal distribution. In this case, the peak value or mode is on the left with a long tail on the right. The mean and median are not very far apart (4 and 4.5) but the mode is much smaller with a value of 3.

ANSWER KEY

1. A	21. B
2. B	22. A
3. D	23. D
4. C	24. C
5. A	25. C
6. D	26. B
7. D	27. B
8. C	28. C
9. C	29. C
10. C	30. C
11. A	31. D
12. C	32. C
13. C	33. C
14. B	34. A
15. D	35. C
16. B	36. A
17. C	37. B
18. C	38. A
19. D	39. B
20. A	40. A

RIGOR TABLE

	Easy 20%	Average 40%	Rigorous 40%
Question	5, 10, 16, 21, 23, 32, 33, 38	1, 2, 4, 7, 8, 11, 14, 17, 25, 26, 29, 30, 34, 36, 37, 39	3, 6, 9, 12, 13, 15, 18, 19, 20, 22, 24, 27, 28, 31, 35, 40